Bibliographic information published by the German National Library:

The German National Library lists this publication in the National Bibliography; detailed bibliographic data are available on the Internet at http://dnb.dnb.de .

Imprint:

Copyright © 2016 GRIN Verlag, Open Publishing GmbH
Print and binding: Books on Demand GmbH, Norderstedt Germany
ISBN: 9783668444294

This book at GRIN:

http://www.grin.com/en/e-book/359220/antibacterial-and-antifungal-properties-of-brahmi

Prem Jose Vazhacharickal et al.

Antibacterial and Antifungal Properties of Brahmi

GRIN Publishing

Studies on antibacterial and antifungal properties of Brahmi (*Bacopa monnieri* (Linn) Pennell)

Prem Jose Vazhacharickal, Jiby John Mathew, Sajeshkumar N.K, Sherin Peter and Riny Susan Thomas

ACKNOWLEDGEMENTS

Firstly we thank **God Almighty** whose blessing were always with us and helped us to complete this project work successfully.

We wish to thank our beloved Manager **Rev. Fr. Dr. George Njarakunnel,** Respected Principal **Dr. Joseph V. J,** Vice Principal **Fr. Joseph Allencheril,** Bursar **Shaji Augustine** and the Management for providing all the necessary facilities in carrying out the study. We express our sincere thanks to **Mr. Binoy A Mulanthra** (lab in charge, Department of Biotechnology) for the support. This research work will not be possible with the co-operation of many farmers.

We are gratefully indebted to our teachers, parents, siblings and friends who were there always for helping us in this project.

Prem Jose Vazhacharickal*, Jiby John Mathew, Sajeshkumar N.K, Sherin Peter and Riny Susan Thomas

Table of contents

Table of figures

Table of tables

List of abbreviations

%	: Percentage
°C	: Centigrade
E.coli	: *Escherichia coli*
g	: Gram
L	: Liter
Mg	: Milligram
MHA	: Mueller hinton agar
mm	: millimeter
NA	: Nutirent agar
NB	: Nutrient broth
PDA	: Potato dextrose agar
µl	: Micro litre

Studies on antibacterial and antifungal properties of Brahmi (*Bacopa monnieri* (Linn) Pennell)

Prem Jose Vazhacharickal[1]*, Jiby John Mathew[1], Sajeshkumar N.K[1], Sherin Peter[1] and Riny Susan Thomas[1]

[1]Department of Biotechnology

Abstract

Fungal and bacterial infections have increased rapidly and the wide uses of synthetic medicines have cytotoxicity on host and made drug resistance among the pathogen. The Antifungal and antibacterial analysis of Bacopa monnieri (Linn) Pennell (Brahmi) was conducted in the present investigation. Various extracts (water, methanol, acetone, petroleum ether and chloroform) of dried Brahmi leaves and stem were tested against two strains of fungi- *Aspergillus niger* and *Candida albicans* as well as bacterial strains. The antibacterial and antifungal activity of different plant extracts was determined by agar well-diffusion method using Muller hinton agar and Sabouraud dextrose agar. The methanol extracts of Brahmi leaves shows inhibition zones on *Aspergillus niger* (12.3 ± 0.6), *Candida albicans* (12.3 ± 0.6), Staphylococcus species (12.3 ± 0.6) and Bacillus species (12.3 ± 0.6). Water extract does not seem to have any good antimicrobial activity against all above mentioned the test microorganisms. The present in vitro investigation results shows that the extracts of Brahmi leaves and stems show good antifungal and antibacterial activity. The study also concludes that methanol and acetone extracts showed good higher efficacy of the bioactive compounds.

1. Introduction

Invasive fungal infections have increased in frequency and severity over the last two decades as a result of an increasing number of immunocompromised hosts [1]. Widespread use of antibacterial and antifungal therapies for curative and prophylactic purposes has serious drawbacks such as the development of fungal resistance and toxic side effects. Because of the eukaryotic properties of fungi, many antifungal compounds exhibit a potent cytotoxic effect on humans which is a significant limitation for the application of these compounds as a practical drug [2]. In search of new antifungal compounds with low side effects such as cytotoxicity, recent efforts have focused on natural resources to obtain a novel bioactive substance, especially from plants. Due to the high costs, adulterations and possible side effects of synthetic drugs the search for alternative cheap medicinal plants is of top priority in developing and under developed countries [3].

Medicinal plants represent a rich source of antimicrobial agents and used widely in Ayurvedic and other traditional medicinal systems [3, 4, 5, 6]. A wide range of medicinal plant parts used for extract as raw drugs and they possess varied medicinal properties. The different parts used include root, stem, flower, fruit, twigs exudates and modified plant organs. While some of these raw drugs are collected in smaller quantities by the local communities and folk healers for local used, many other raw drugs are collected in larger quantities and traded in the market as the raw material for many herbal industries [7]. Although hundreds of plant species have been tested for antimicrobial properties, the vast majority of have not been adequately evaluated [8].

The common medicinal plant *Bacopa monnieri* (Linn) Pennell, is a well-known herb in Indian system of medicine commonly called as Brahmi. The plant is commonly found in wet, damp and marshy areas. The active ingredients of the plants were used as brain tonic, which is effective in maintaining the vigour and intellect [9, 10] contributed by a triterpenoid saponin called bacosides [11, 12]. The bacosides enhance the efficiency of transmission of nerve impulse and also used as a laxative and curative for ulcers, inflammation, anaemia, scabies, leucoderma, epilepsy and asthma [13, 14]. The plant is also reported to show sedative [15], hyperthyroidism, vasoconstrictor and anti-inflammatory property [16].

1.1 Objectives

Given lacking qualitative and quantitative data on antimicrobial and antifungal properties *Bacopa monnieri* (Linn) Pennell plant extracts, our objectives were to (1) Determination of the extraction of Brahmi using various extractants and (2) validation of the various extractants using zone of inhibition studies on major bacterial and fungal strains causing food spoilage and food borne diseases.

1.2 Scope of the study

The study using the extracts of *Bacopa monnieri* (Linn) Pennell could develop new drug formulations to combat various diseases. Moreover such studies could improve the potential value of the underutilized plants especially *Bacopa monnieri* (Linn) Pennell.

1.3 Taxonomical classification

Kingdom: Plantae-- planta, plantes, plants, vegetal

Subkingdom: Viridiplantae

Division: Tracheophyta – vascular plants, tracheophytes

Class: Magnoliopsida -- dicots, dicotyledones, dicotyledons

Order: Lamiales

Family: Plantaginaceae – plantains

Genus: Bacopa Aubl.

Species: *Bacopa monnieri* (Linn) Pennell

Table 1. Different vernacular names of *Annona reticulata* around the globe and India.

Language	Names
Scientific names	*Bacopa monnieri* (Linn) Pennell
Name in various global languages	
English	Indian pennywort
Name in various Indian languages	
Sanskrit	Brahmi
Hindi	Brahmi
Bengali	Brahmisaka
Marathi	Brahmi
Kannada	Brahmi
Konkani	Brahmi
Malayalam	Brahmi/Kodakan/Muthil
Tamil	Nirp-pirami

3. Hypothesis

The current research work is based on the following hypothesis

1) The *Bacopa monnieri* (Linn) Pennell extracts vary in their antimicrobial and antifungal activities.

4. Materials and Methods

4.1 Study area

Kerala state covers an area of 38,863 km^2 with a population density of 859 per km^2 and spread across 14 districts. The climate is characterized by tropical wet and dry with average annual rainfall amounts to 2,817 ± 406 mm and mean annual temperature is 26.8°C (averages from 1871-2005; [19]). Maximum rainfall occurs from June to September mainly due to South West Monsoon and temperatures are highest in May and November (Figure 1).

4.2 Plant sample collection

Fresh plant materials of *Bacopa monnieri* (Linn) Pennell (Brahmi) was collected from the herbal garden of Mar Augusthinose college, Kerala, India (Figure 2). The herbal garden was established on 2006 as a central government scheme form Medicinal Plant Board to educate the students about the local medicinal plants. The garden has a size of 0.1 ha with more than 200 species of medicinal plants which were mainly maintained by the students. The leaves and stems were collected separately. The plant materials were thoroughly washed with distilled water and fresh weight were determined. The samples are then oven dried (KOA4, KEMI lab equipments, Ernakulam, India) at 60°C for 24 h. The dried samples were powdered using a waring blender (Magic V2, Preethi Kitchen Appliances Pvt Ltd, Chennai, India) and stored in air-tight polyethylene bottles until further analysis.

4.3 Preparation of extract

Ten gram of Brahmi plant powder was mixed with 50 ml of different solvents like water, methanol, acetone, petroleum ether and chloroform as extractant. The solution was agitated for 24 h supplying constant heat at 60°C using a shaking incubator (KHM4, KEMI lab equipments, Ernakulam, India). The final extractants were filtered using a Whatmann filter paper 42 (GE Healthcare UK Ltd, Buckinghamshire, UK) and transferred to an airtight polyethylene bottles which were stored on refrigerator for further inhibition studies.

4.4 Microorganism

Bacterial strains were obtained from bacterial stock, Department of Biotechnology, Mar Augusthinose College, Ramapuram, India which was previously collected from Microbial Type Culture Collection, Chandigarh, India. The selected bacterial strain includes *Staphylococcus species, Bacillus species* while fungal strain includes *Aspergillus niger* and *Candida albicans*. The bacterial cultures were maintained at 4°C on nutrient agar slants while the fungal strains were maintained in Sabouraud Dextrose Agar slants.

4.5 Agar well diffusion tests

The antibacterial and antifungal activity of different plant extracts was determined by agar well-diffusion method according to Ahmed et al. For this, 150 µl of 12-16 hrs incubated cultures of bacterial species were mixed in molten Muller Hinton Agar (MHA) medium (Himedia Laboratories, Mumbai, India) and poured in pre-sterilized petri plates. A well puncher of six mm diameter (Hindustan Syringes & Medical Devices Ltd, New Delhi, India) used to punch wells in solidified medium and later filled with extracts. The only difference in the antifungal assay is the use of Sabouraud Dextrose Agar instead of MHA. Five treatments were used including 50 µl of plant extract in water, plant extract in acetone, chloroform, methanol and petroleum ether. The treatments were done separately of leaves and stem extracts of Brahmi. All the treatments were replicated thrice to avoid possible errors and misinterpretations. Additionally control wells were also made which were loaded with the extracatnt alone. The plates were incubated at 37°C for 24 hrs in incubator and the diameter of the zone of inhibition was measured and recorded. The antibacterial and antifungal activity was interpreted from the size of the diameter of zone of inhibition measured to the nearest mm as observed from the clear zones surrounding the wells.

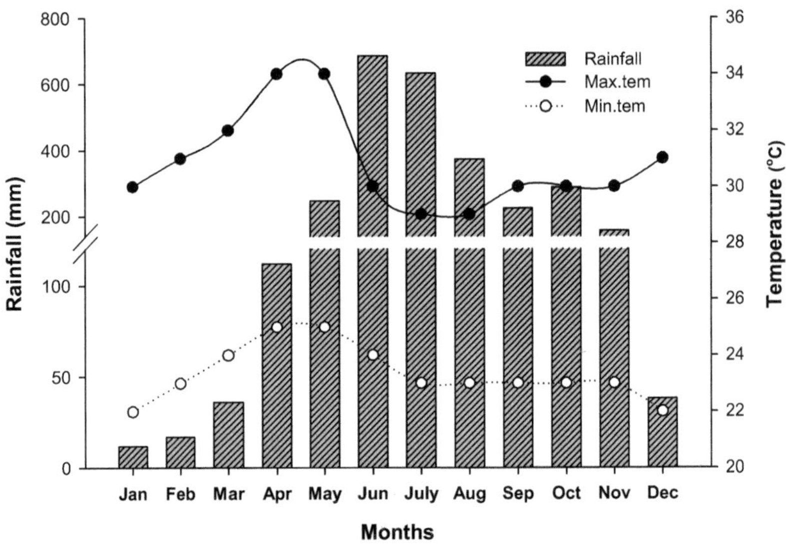

Figure 1. Mean monthly rainfall (mm), maximum and minimum temperatures (°C) in Kerala, India (1871-2005; [19]).

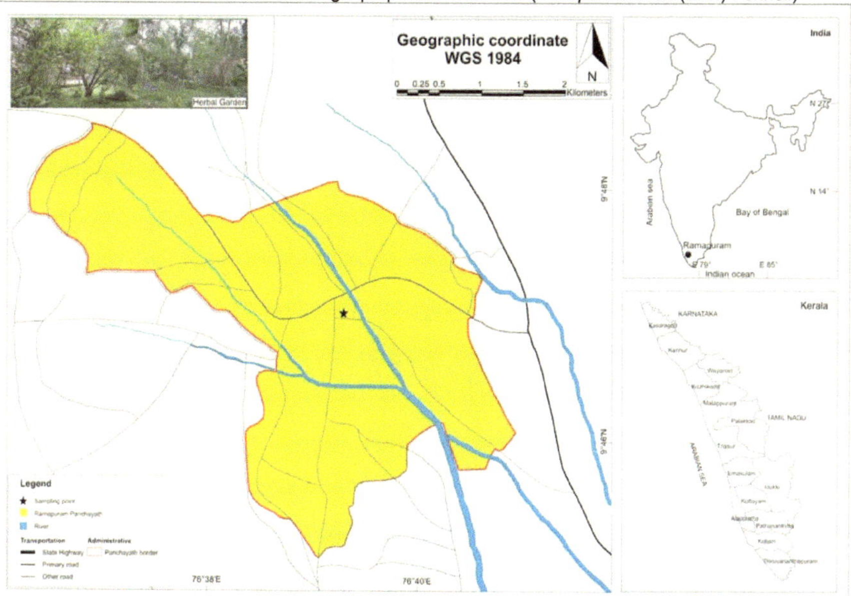

Figure 2. Map of Ramapuram showing the location of the sample collection point.

4.8 Statistical analysis

Descriptive statistics using SPSS 12.0 (SPSS Inc., Chicago, IL, USA) were conducted to summarize the data.

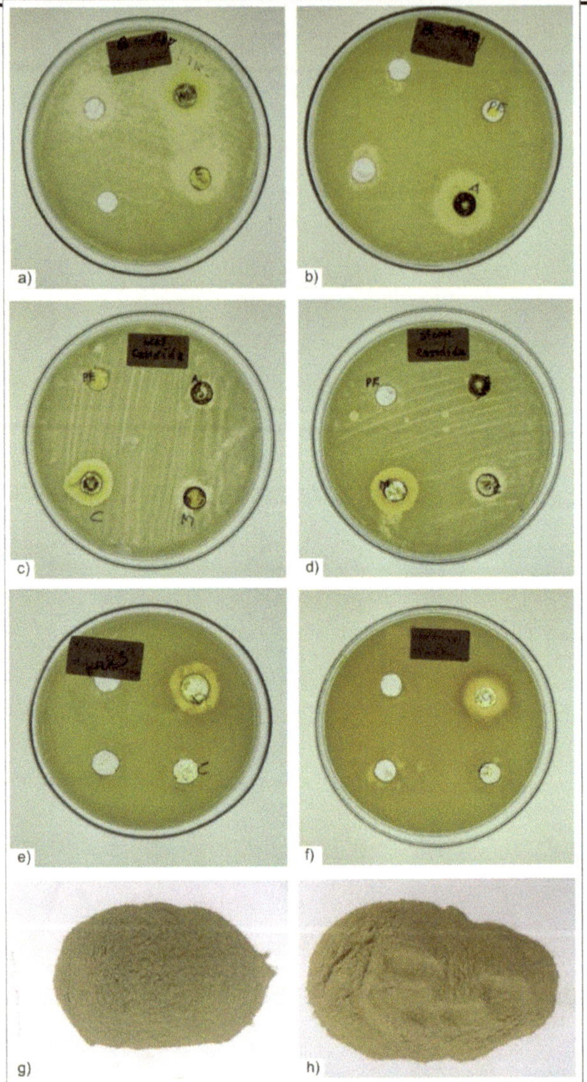

Figure 3. Inhibition zone produced by *Bacopa monnieri* (Linn) Pennell (Brahmi) on different bacterial and fungal strains (leaf and stem respectively), a) and b) *Aspergillus niger,* c) and d) *Candida albicans,* e) and f) *Staphylococcus species,* g) Brahmi leaves powder, h) Brahmi stem powder.

Table 2. Inhibition zone of various control treatments (petroleum ether, chloroform, acetone, methanol and water) against various microorganisms using Muller Hinton Agar and Sabouraud Dextrose Agar (n=3, values in mm).

Microorganism	Controls				
	Petroleum ether	Chloroform	Acetone	Methanol	Water
Staphylococcus spp.	1.2± 0.3	1.0± 0.3	0.9 ± 0.3	1.2± 0.3	-
Bacillus spp.	1.2 ± 0.3	0.9 ± 0.3	1.2 ± 0.6	1.2 ± 0.6	-
Aspergillus niger	0.6 ± 0.3	0.9 ± 0.3	0.9 ± 0.3	0.6 ± 0.3	-
Candida albicans	0.6 ± 0.3	0.6 ± 0.3	0.6 ± 0.3	0.6 ± 0.3	-

Numbers represent means ± one standard deviation (SD) of the mean.

-: no inhibition

Table 3. Inhibition zone of *Bacopa monnieri* (Linn) Pennell (Brahmi) leaves treatments (petroleum ether, chloroform, acetone, methanol and water) against various microorganisms using Muller Hinton Agar and Sabouraud Dextrose Agar (n=3, values in mm).

Microorganism	Treatments				
	Petroleum ether	Chloroform	Acetone	Methanol	Water
Staphylococcus spp.	10.2± 0.6	1.0± 0.3	13.9 ± 0.3	11.2± 0.5	2.8 ± 0.6
Bacillus spp.	1.2 ± 0.3	0.9 ± 0.3	1.2 ± 0.6	12.2 ± 0.6	-
Aspergillus niger	0.6 ± 0.3	21.4 ± 0.8	21.6 ± 0.6	24.0 ± 0.5	-
Candida albicans	0.6 ± 0.3	14.1 ± 0.3	11.6 ± 0.6	19.1 ± 0.2	-

Numbers represent means ± one standard deviation (SD) of the mean.

-: no inhibition

Table 4. Inhibition zone of *Bacopa monnieri* (Linn) Pennell (Brahmi) shoot treatments (petroleum ether, chloroform, acetone, methanol and water) against various microorganisms using Muller Hinton Agar and Sabouraud Dextrose Agar (n=3, values in mm).

Microorganism	Treatments				
	Petroleum ether	Chloroform	Acetone	Methanol	Water
Staphylococcus spp.	1.2± 0.3	1.0± 0.3	0.9 ± 0.3	1.2± 0.3	-
Bacillus spp.	1.2 ± 0.3	0.9 ± 0.3	1.2 ± 0.6	1.2 ± 0.6	-
Aspergillus niger	0.6 ± 0.3	0.9 ± 0.3	0.9 ± 0.3	0.6 ± 0.3	-
Candida albicans	15.2 ± 0.3	13.6 ± 0.5	12.6 ± 0.3	0.6 ± 0.3	-

Numbers represent means ± one standard deviation (SD) of the mean.

-: no inhibition

5. Results

From the zone of Inhibition measurements, water extract does not seem to have any good antimicrobial activity against all above mentioned the test microorganisms (Table 3 and 4). The methanol extracts of Brahmi leaves shows inhibition zones on *Aspergillus niger* (12.3 ± 0.6), *Candida albicans* (12.3 ± 0.6), *Staphylococcus species* (12.3 ± 0.6) and *Bacillus species* (12.3 ± 0.6). Surprisingly, the methanol extract of Brahmi stem showed inhibition only to *Candida albicans* (12.3 ± 0.6).

Similar trends were observed with acetone extract of Brahmi leaves with clear zone of inhibition against *Aspergillus niger* (12.3 ± 0.6), *Candida albicans* (12.3 ± 0.6), *Staphylococcus species.* Petroleum ether extract of Bhrami leaves showed zone of inhibition only against *Staphylococcus species* while the chloroform extracts showed inhibition against *Aspergillus niger* (Figure 3; Table 4).

Brahmi stem extract does not show any prominent antimicrobial activity against any of the test microbial strains except *Candida albicans*. The methanol, acetone and chloroform extract shows good antifungal activity for *Candida albicans*.

5. Discussion

Antibiotic resistance among bacterial is of high concern in clinical areas. The selected bacterial strains are a common cause for food poisoning, food spoilage and many other gastrointestinal troubles and have the possibility to become antibiotic resistant. The present study shows the improvement of antibacterial activity of honey with different plant extracts.

The active ingredients which supply a plant antimicrobial property is a part of the natural defence system and may be contributed by the secondary metabolites including alkaloids and other phenolic compounds [17]. The solubility of these compounds may vary with the solvents used [18].

The extracts Brahmi with water seems to be less capable of explicating its antimicrobial properties makes the importance of various extract solvents. More over slight difference antimicrobial properties regarding the type of plant were also observed indicating that the active ingredients are partitioned at various tissues in variable amounts.

In general methanol extracts of Brahmi leaves shows good antimicrobial and antifungal activities against the test microorganisms. The antibacterial activities of leaves with other extracts except water also show high variations. Acetone and chloroform seemed to better solvents especially for test bacterial and fungal strains. Interestingly Brahmi shoot extracts with acetone, chloroform and petroleum ether shows promising antifungal property against *Candida albicans*.

6. Conclusions

The present in vitro investigation results shows that the extracts of Brahmi leaves and stems show good antifungal and antibacterial activity. The study also concludes that methanol and acetone extracts showed good higher efficacy of the bioactive compounds. Further study should be also conducted to reduce the research gaps in this area.

Conflict of interest statement

We declare that we have no conflict of interest.

Acknowledgements

The authors are grateful for the cooperation of the management of Mar Augusthinose college for necessary support. Technical support from Binoy A Mulanthra is also acknowledged.

References

[1] Tortorano AM, Caspani L, Rigoni AL, Biraghi E, Sicignano A, and Viviani MA. Candidosis in the intensive care unit: a 20-year survey. J Hosp Infect. 2004; 5:8-13.

[2] Singh HK, Dhawan BN. Neuropsychopharmacological effects of the Ayurvedic nootropic Bacopa monniera Linn.(Brahmi). Indian J Pharmacol. 1997; 29(5):359-65.

[3] Arora DS, Kaur GJ. Antibacterial activity of some Indian medicinal plants. J Nat Med. 2007; 61(3):313-7.

[4] Sampathkumar P, Dheeba B, Vidhyasagar ZV, Arulprakash T, Vinothkannan R. Potential antimicrobial activity of various extracts of Bacopa monnieri (Linn.). Int J Pharmacol. 2008; 4(3): 230-2.

[5] Thomson WAR. Medicines from the earth: a guide to healing plants. New York: McGraw-Hill, 1978; p 25-9.

[6] Mahesh B, Satish S. Antimicrobial activity of some important medicinal plant against plant and human pathogens. World journal of agricultural sciences. 2008; 4(5):839-43.

[7] Uniyal SK, Singh KN, Jamwal P, Lal B. Traditional use of medicinal plants among the tribal communities of Chhota Bhangal, Western Himalaya. J Ethnobiol and Ethnomed. 2006; 2(1): 2-14.

[8] Balandrin M F, Klocke, JA, Wurtele ES, Bollinger WH. Natural plant chemicals: sources of industrial and medicinal materials. Sci. 1985; 228:1154-9.

[9] Ahmad I, Mehmood Z, Mohammad F. Screening of some Indian medicinal plants for their antimicrobial properties. J Ethnopharmacol. 1998; 62(2):183-93.

[10]   Mehrotra BN. Compendium of Indian medicinal plants Lucknow: Central Drug Research Institute, 1990; p: 226-7.

[11]   Chaudhuri PK, Srivastava R, Kumar S, Kumar S. Phytotoxic and antimicrobial constituents of Bacopa monnieri and Holmskioldia sanguinea. Phytother Res. 2004; 18(2):114-7.

[12]   Ghosh T, Maity TK, Bose A, Dash G.K, Das M. Antimicrobial activity of various fractions of ethanol extract of *Bacopa monnieri* Linn. aerial parts. Indian J Phar Sci. 2007; 69(2): 312-4.

[13]   Russo A, Borrelli F. *Bacopa monniera*, a reputed nootropic plant: an overview. Phytomedicine. 2005; 12(4):305-17.

[14]   Gohil KJ, Patel JA. A review on *Bacopa monniera*: Current research and future prospects. Int J Gre Pha. 2010; 4(1): 1-9.

[15]   Kar A, Panda S, Bharti S. Relative efficacy of three medicinal plant extracts in the alteration of thyroid hormone concentrations in male mice. J Ethnopharmacol. 2002; 81(2):281-5.

[16]   Khan AV, Ahmed QU, Shukla I, Khan AA. Antibacterial efficacy of *Bacopa monnieri* leaf extracts against pathogenic bacteria. Asian Biomed. 2010; 4(4):651-5.

[17]   Robinson T. Metabolism and function of alkaloids in plants. Sci. 1974; 184:430-5.

[18]   Huie CW. A review of modern sample-preparation techniques for the extraction and analysis of medicinal plants. Anal Bioanal Chem. 2002; 373:23-30.

[19]   Krishnakumar, K. N., GSLHV Prasada Rao, and C. S. Gopakumar. "Rainfall trends in twentieth century over Kerala, India." *Atmospheric environment*43.11 (2009): 1940-1944.

YOUR KNOWLEDGE HAS VALUE

- We will publish your bachelor's and master's thesis, essays and papers

- Your own eBook and book - sold worldwide in all relevant shops

- Earn money with each sale

Upload your text at www.GRIN.com and publish for free